MW01114174

Isabella's Peppermint Flowers

by Susan Leopold

illustrated by Nicky Staunton

This book is dedicated to Marion Lobstein, botany professor extraordinaire, whose support, encouragement, and scientific review was critical to this project. Sad but true, botany departments are closing their doors and children are learning less and less about the historical significance of botany and the critical role that plants play in sustaining life on earth.

We are in the midst of the sixth mass extinction, with plant species disappearing at alarming rates. We need botanists! We need young people to embrace the wonders of plant life and to be ambassadors for the ancient beings that make life possible on this planet we call home.

Proceeds from this book will go to the Foundation of the Flora of Virginia Project to support educational endeavors that inspire children to embrace the study of botany and to learn about Virginia's ecological diversity!

ISBN: 978-0-692-33302-0

Rev. date: 11/14/2014

To order additional copies of this book, visit: www.FloraForKids.org
Graphic design work for this book done by: www.WeathervaneGraphics.com

It was springtime, and Isabella, who lived on a farm in Virginia, wanted to go flower hunting in the woods with her younger sister, Flora May, and their mother. Isabella was quite insistent, for she knew about the **spring ephemerals**, those woodland wildflowers that appeared for only a short, magical time. She had learned that, just before the trees' leaves reappeared and reformed their dense canopy, the spring ephemerals would start to emerge.

The warmer days, enhanced by the sunlight that heated the forest floor, signaled the spring ephemerals that it was time. Isabella did not want to miss seeing those radiant flowers that, not long after they showed themselves, would disappear again and then go dormant till the next spring.

"Mom, can we go flower hunting?" Isabella asked.

"That's a wonderful idea," their mother answered.

Since many spring ephemerals are **rare** and some are **endangered**, their mother reminded the girls to take their camera to photograph the woodland wildflowers they encountered. Then they could use the pictures in their nature notebooks instead of the flowers themselves.

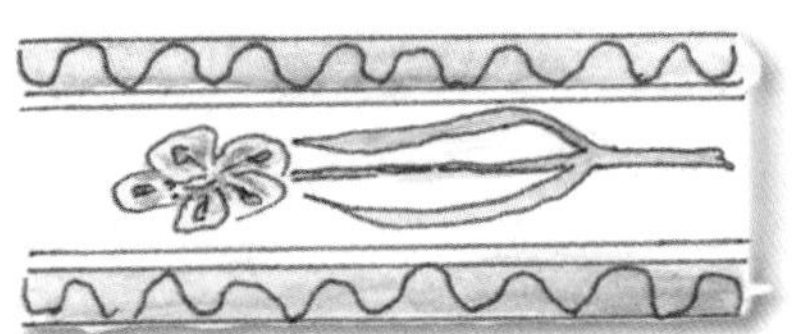

"Okay, let's get going!" Isabella said. "Shall we go to our favorite spot?"

"The peppermint-candy patch!" Flora May shouted.

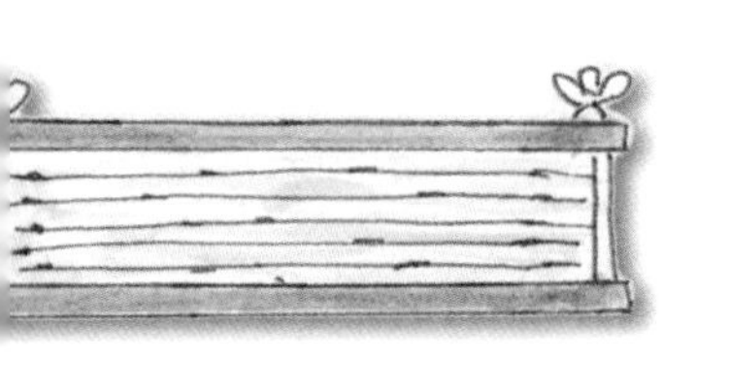

It wasn't really a patch of peppermint candy. But deep in the forest was a carpet of small white flowers with pink stripes that reminded the girls of peppermint candy. It was their favorite place.

When the girls first discovered the patch of pink-and-white flowers, their mother had told them they were called **spring beauties**.

"Tell us a story while we walk!" Isabella said, buttoning Flora May's jacket. The girls put on rubber boots so they would be prepared for the wet ground along Fiery Run, the shallow stream they would follow in search of the spring beauties.

"Well," began their mother, "all plants are special, and each one has a story to tell!" And with that, they set off into the woods.

"Are you listening?" Isabella asked little Flora May. Flora May was already stooping down to peer at the leaves, pine needles, and tiny flowers on the forest floor.

"Flora May doesn't need to listen," their mother said reassuringly. "She's concentrating."

"But she'll pick every flower in sight, Mom," Isabella protested.

"Hush," their mother said quietly. "You and I can keep a close eye on her." Then, to both, she said, "Have a seat and I'll tell you how the candystriped spring beauty got its name!"

Isabella and Flora May sat on a log that was covered with lichens, fungi, and moss. Their mother explained that the story of the spring beauty reminds us of how the first book on Virginia plants came to be written, by a man named **John Clayton** in the 1700s. It was called *Flora Virginica*.

"*Flora?*" said the littler sister, suddenly alert. "That's my name!" she exclaimed.

Their mother nodded. "The word 'flora' also refers to the plants that are found in a specific place, like the state of Virginia. And a book about the plants of a place is also called a flora."

"So Flora's name means more than just flowers!" Isabella said.

"That's right," their mother said. She went on to tell the girls how a flora is a book that can help someone figure out the name of an unknown plant.

"A flora describes plants and their flowers in great detail," she said. "When you find a plant you don't know, you can use a flora's keys to find out what kind of plant you have found."

"At first glance, different plants can look a lot alike," their mother said. "A flora has **dichotomous keys** that will guide you through a series of steps that help you identify your plant." At each step, she explained, you read two phrases that describe something about the plant. In each pair, one phrase does not quite describe your plant, but the other one does. You may have to go through two, or three, or even more steps, but eventually, you reach the last pair of phrases. "Again, one will be wrong," she said. "The right one will give you the name of your plant!"

Their mother explained that John Clayton had described the flowers, ferns, lichens, algae, conifers, and even fungi so that **botanists** and other scientists in Europe could better understand the plants from the **colony of Virginia**.

"Virginia is where we live!" said Flora May.

"That's right," their mother replied. "The first flora of the New World, *Flora Virginica*, was big news back in England, and this book made John Clayton famous. Thomas Jefferson, our third president, called him America's finest botanist.[1] That was quite a compliment, coming from a plant lover like Jefferson!"

Clayton was born in England in 1694. His father came to Virginia and was attorney general of the colony, but young John stayed in England to complete his education.

1 Jefferson, Thomas. *Notes on the State of Virginia.* London, 1787.

When he was just 21, he came to Virginia. This was an exciting time for Clayton, because as a boy he had learned to appreciate gardening, and in Virginia he encountered many plants he had never seen before. These plants, of course, were also new to his countrymen back home in England, who loved to study and grow exotic plants in their gardens and greenhouses.

That sounds just like me, Isabella thought as she tried to imagine what Clayton might have looked like.

As they walked along Fiery Run, their mother told them that if you really cared about something in life, you tended to be inspired by someone you admired in that field. For Clayton, that was **Mark Catesby**. Catesby was a remarkable naturalist, as well as a talented artist and writer. His work captured the imagination of Europeans who wanted to know more about the plants and wildlife of the New World. His illustrations were important because there were no cameras in the 1700s. Catesby had to draw what he was studying, and to do so he had to observe in nature what he wanted to draw.

Tulip-tree
Liriodendron tulipifera

Catesby described not only the plants, birds, fish, reptiles, insects, and animals he saw on his travels but also the American Indians he met. Their knowledge and deep understanding of nature were remarkable. Catesby wrote of his respect for the native people, documenting their skills at navigating the land as they traveled great distances and describing their abilities in hunting and cooking the wild plants and meats of the forest.

Catesby was a careful observer of the natural world. In the introduction of his book *The Natural History of Carolina, Florida and the Bahama Islands*, you can sense his skill and attention to detail as he describes the small changes that take place in the color of plants throughout the seasons of the year:

> *...but what form plants are possessed of at different times of the year, and the same plant changes its color gradually with its age: for in spring, the woods and all plants in general are more yellow and bright and as the summer advances, the greens grow deeper, and the nearer their fall, are yet of a more dark and dirty color. What I infer from this, that by comparing a painting with a living plant, the difference of color, if any, is from the above mentioned cause.*[2]

The girls' mother explained that not only were there no cameras, but there were no phones or e-mail either, and that Catesby hand-wrote letters to his friends in Europe while he traveled. In one letter he described the first time Clayton had come looking for him.

> *...he came riding up a beautiful young* **Apollo** *with piercing blue eyes.... he wanted to know more about plants! He told me things I did not know, how the Indians and county people used plants for curing ills and dyeing cloth.*[3]

For Clayton, this encounter was very important because thereafter he and Catesby would exchange tips on how to prepare seeds and roots for shipping to Europe. Catesby also showed Clayton how to dry the plants he collected so that they would make good herbarium specimens. A herbarium specimen is a dried plant that is carefully stitched or glued to a sheet of paper so that its stems, leaves and flower parts, such as the sepals, petals, pistils and stamens, can be easily seen. A label bearing the name of the plant, where and when it was collected, and the name of the collector accompanies each specimen. A unique traveling case called a **vasculum** was often used for carrying plants collected on plant-hunting excursions, so that they could be well protected.

---

2 Catesby, Mark. *The Natural History of Carolina, Florida and the Bahama Islands*. London, 1771.
3 Frye, Harriet. *The Great Forest, John Clayton, and Flora*. Dragon Run Books, Hampton, 1990.

APRIL 15 1740
GLOUCESTER VA
FLOOD PLAIN
OPPOSITE LEAVES
5 PETALS
WHITE-PINK
JOHN CLAYTON

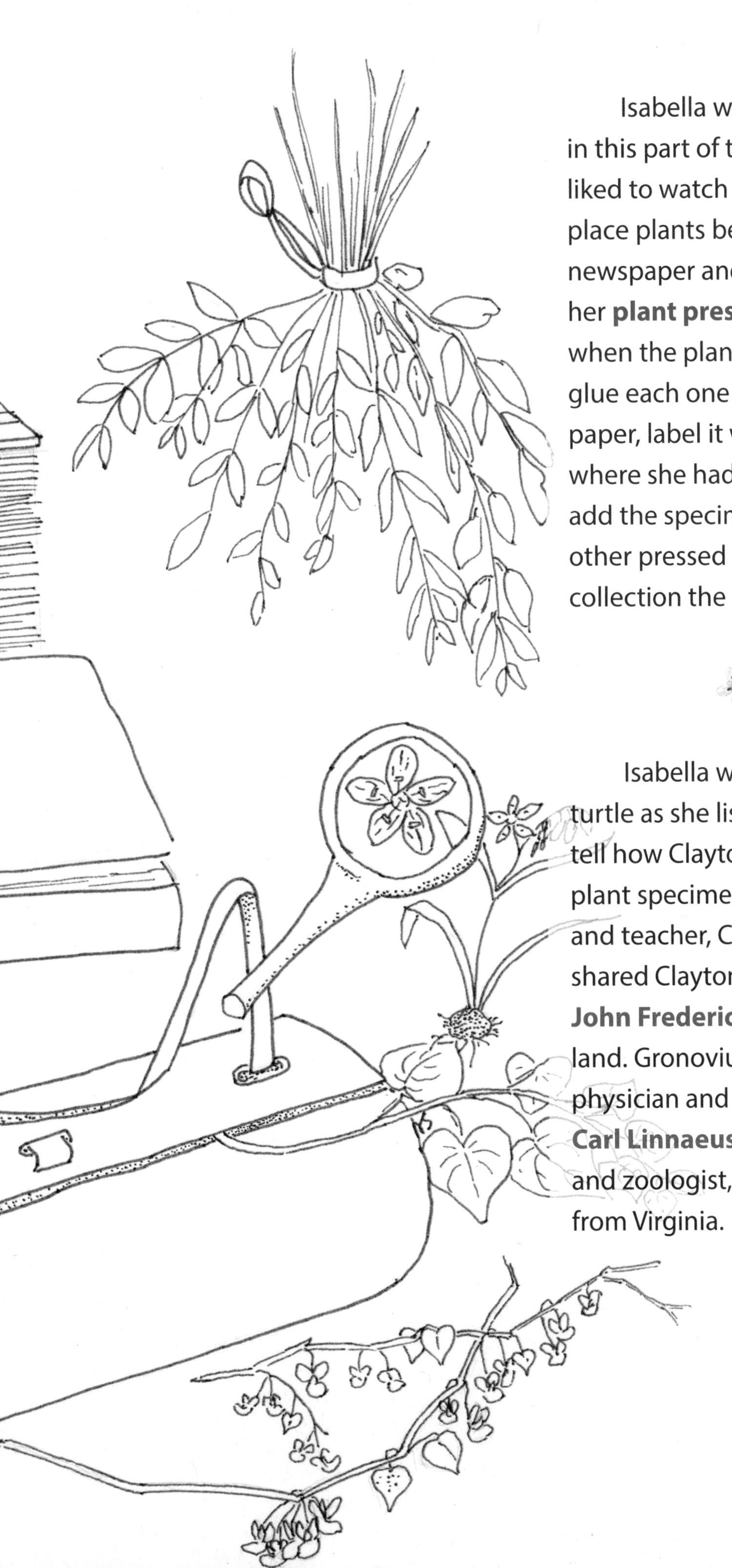

Isabella was especially interested in this part of the story because she liked to watch her mother carefully place plants between two pieces of newspaper and then put them into her **plant press**. After a few days, when the plants had dried, she would glue each one to a sheet of white paper, label it with information about where she had collected it, and then add the specimen to her collection of other pressed plants. She called this collection the Fiery Run Herbarium.

Isabella was watching a box turtle as she listened to her mother tell how Clayton decided to ship his plant specimens abroad to his friend and teacher, Catesby. Catesby then shared Clayton's plants with his friend **John Frederick Gronovius** in Holland. Gronovius, who was a scientist, physician and senator, then asked **Carl Linnaeus**, a botanist, physician, and zoologist, to study the plants from Virginia.

Linnaeus would come to be called the "prince of botany," because he devised **binomial nomenclature** – a fancy way of saying that each species would get two names. The first name identifies the **genus** to which the plant belongs, and its first letter is always capitalized; the second name identifies the species itself, and its first letter is not capitalized.

Linnaeus named many plant species during this time of botanical exploration, and the names he gave plants often honored the botanists who discovered them or the places where they had collected the specimens. When Linnaeus named the spring beauty, he called it *Claytonia virginica*, which recognized Clayton for his contributions to botany and the place, Virginia, where he had collected the specimen.

"So it has many names?" Isabella asked.

"The flower we found in the woods has a common name in English, spring beauty," their mother explained, "but it also has a Latin name, *Claytonia virginica*."

Knowing how much his friends in Europe loved to study new plants, in 1738 Clayton sent Gronovius a manuscript titled *Catalogue of Herbs, Fruits and Trees Native to Virginia*. This catalog included more than 500 plants that Clayton had collected, studied and, in many cases, grown in his garden.

Try to imagine the perilous route that each letter, each seed, each plant specimen, even Clayton's catalog had to travel to reach Europe in those days! At a time when pirates, shipwrecks, and deadly storms threatened a small ship sailing on the North Atlantic Ocean, the way this story ended is quite a miracle. In spite of the dangers, Clayton's catalog made it safely into Gronovius's hands, and Gronovius was so excited that he decided to publish it. The result was the book he named *Flora Virginica*.

At last the sisters reached the patch of peppermint flowers they were looking for. Their mother carefully dug around the base of one of the spring beauties so she could show the girls the round swelling, called a **corm**, at the base of the flower's stem.

"Spring beauties were not hunted just by us or Clayton but also by American Indian and colonial children, who loved to find spring beauty flowers because the corms were a treat to eat."

"The corm may look like a root," their mother told the girls, "but it's really the plant's underground storage, where it stores food so it can lie dormant for the entire summer and winter and then appear for just these brief few weeks each spring. You can see why children once called these corms fairy spuds!" The corms also provide food for chipmunks and other wildlife, she added.

The girls' mother then carefully replanted the corm so the plant could bloom again the next spring.

She told the girls how the pink lines that made them think of peppermint candy served a distinct purpose. Called "nectar lines," they reflect ultraviolet light, which insect pollinators (but not people!) can see and which help guide insects to the flower. The pollinators get nectar, but the flowers are pollinated in the process, and so make seeds that can grow into new plants.

Three Important Pollinators of *Claytonia virginica*

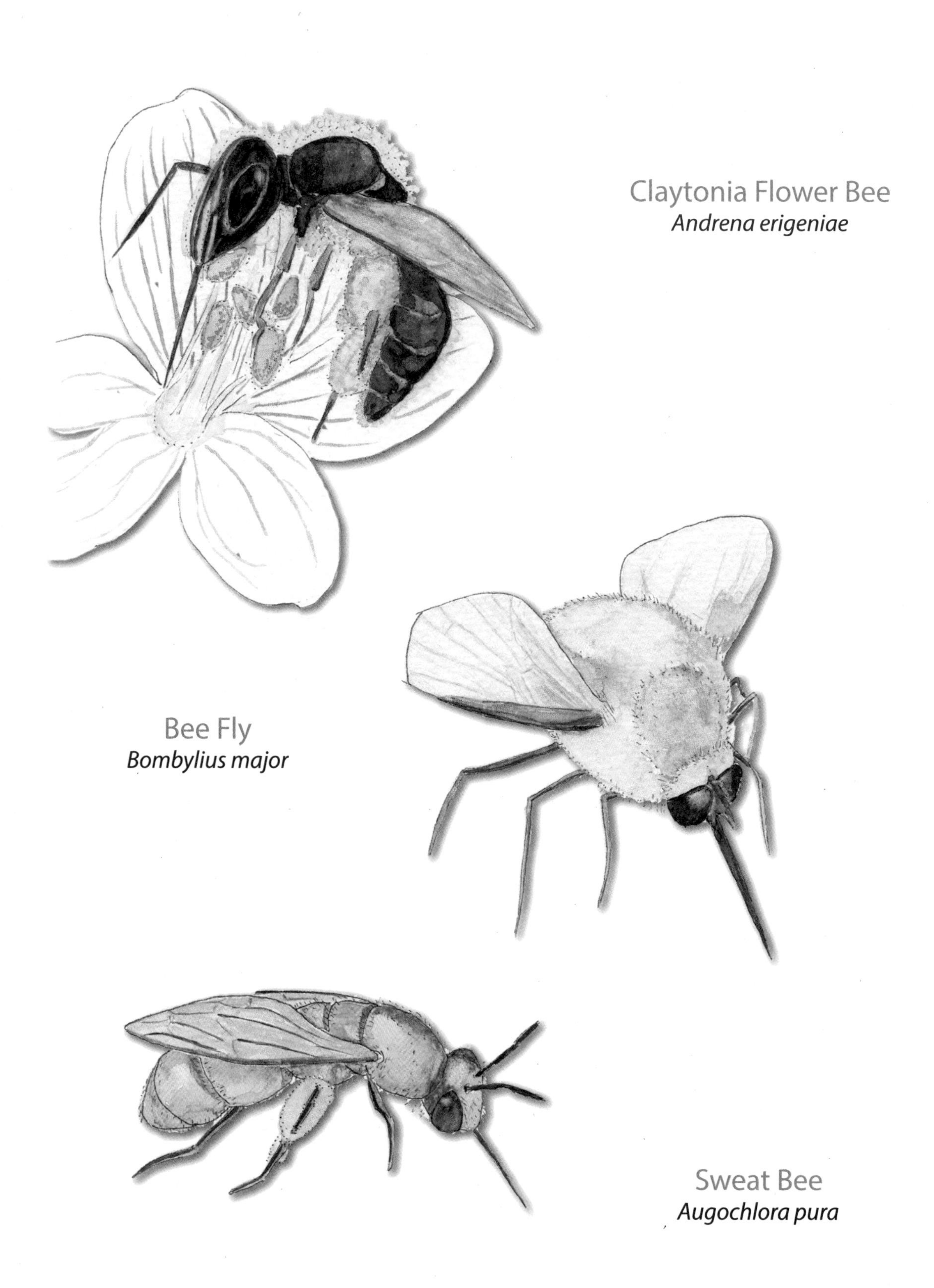

"But it's not just bees that are interested in the spring beauty," their mother told the girls. "When the flower sets seed," she explained, "attached to each seed is a small piece of white material called an **elaiosome**. The elaiosomes are appealing to ants – like delicious marshmallows!"

"Marshmallows for ants?" Flora giggled. "I love marshmallows!"

"You might think elaiosomes would taste sweet just like marshmallows," their mother continued, "but these marshmallows aren't made of sugar – they're made of fat! Fat gives the ants energy in the spring so they can work hard for their colony. What's more, when ants carry away the spring beauty seeds to use the fatty elaiosome as food, they are also dispersing the seeds around the forest. The seeds later germinate and form new plants."

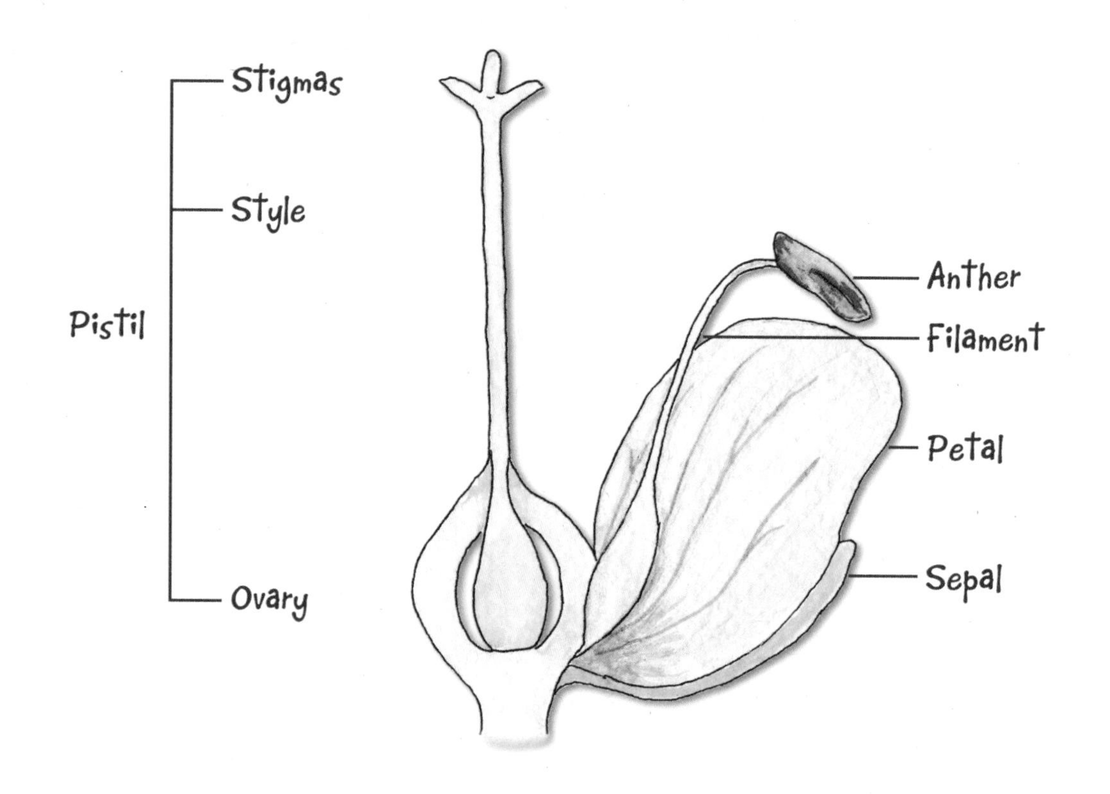
Stigmas
Style
Pistil
Ovary
Anther
Filament
Petal
Sepal

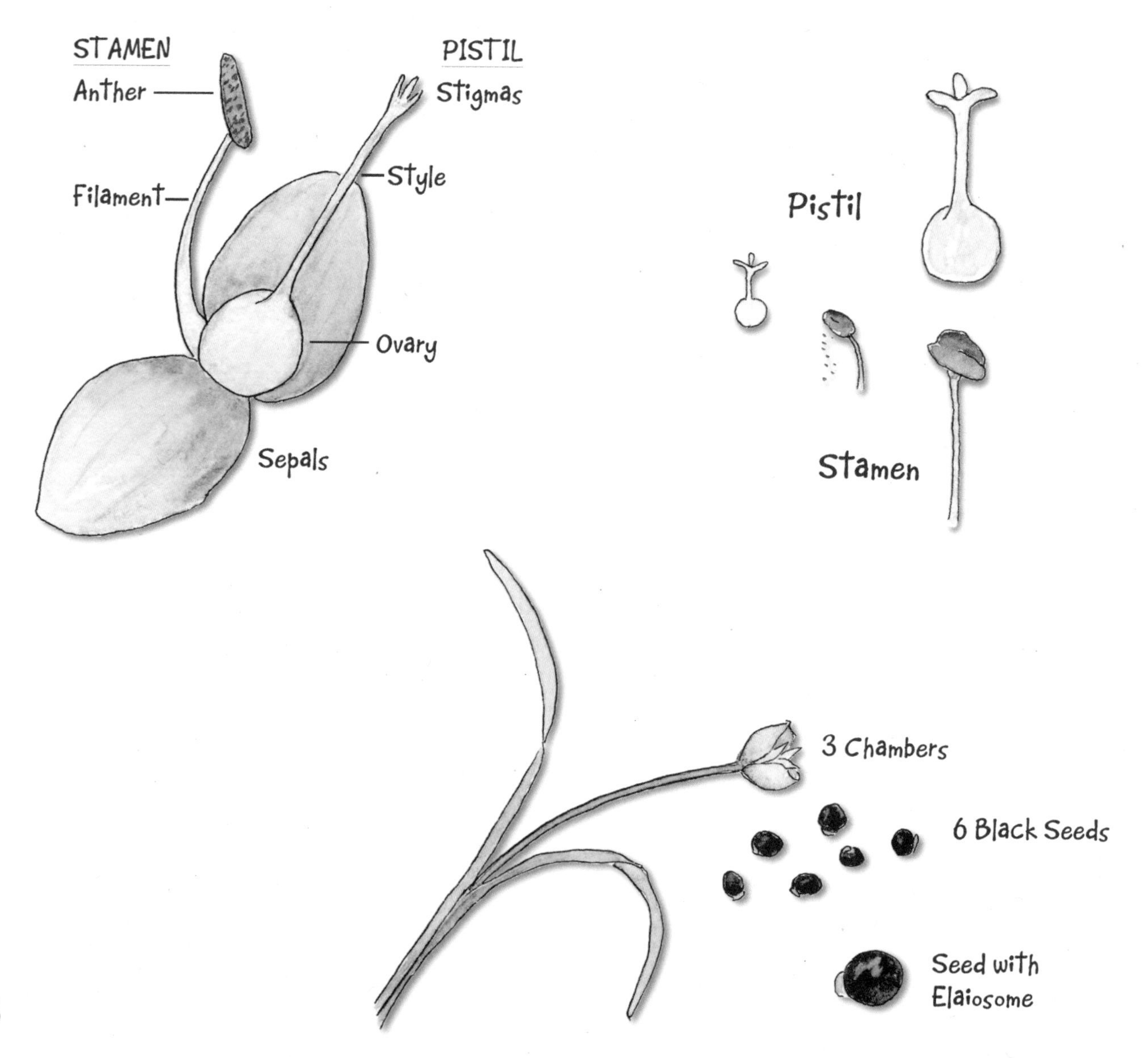
STAMEN
Anther
Filament
PISTIL
Stigmas
Style
Ovary
Sepals
Pistil
Stamen
3 Chambers
6 Black Seeds
Seed with
Elaiosome

As Flora ran from patch to patch pretending to be a bee, their mother told Isabella more about John Clayton.

"He was the clerk of Gloucester County," she said, explaining that he recorded official documents such as court cases, wills, surveys, and deeds. "But still he found time to run his farm and tend his garden. When he could get away, he rode his horse or hiked around Virginia in search of plants that only a handful of people in Europe had ever seen. Clayton also wrote about how the plants were used by Virginia's Indians for medicines, especially to treat snakebite."

“The spring beauty’s flower teaches us about the many parts just one wildflower species plays in the forest **ecosystem**,” their mother reminded the girls. “Chipmunks eat its corms. Bees and other insects are attracted to its nectar, and, as they gather it, they pollinate the plants. Ants are eager to eat the elaiosomes, and in the process they gather the seeds and disperse them around the forest floor.”

On the walk home Isabella thought about the story of John Clayton and the naming of the spring beauty and how one unique plant touched the lives of some very important people who were botanists, doctors, artists and naturalists – all inspired by the wonders of nature.

Flora May wasn’t paying much attention. It seemed too confusing until their mother pointed to a spring beauty flower and reminded her that they were talking about the pink-and-white-striped flowers that they loved so well.

“Our peppermint flowers!” Flora May shouted.

“Our peppermint flowers,” replied Isabella, “the sweetest name of them all!” ■

Definitions

Apollo was a character of Greek mythology and was recognized as a god of light, sun, truth, and prophecy. To compare someone to Apollo is a great compliment.

Binomial nomenclature is the scientific naming of specific types of organisms by assigning each a two-part or a binomial name. The first part of each name is the genus (plural: genera) and the second is the species name.

Botanist is a person who studies plants. The history of botany is rooted in the medicinal and edible uses of plants as well as in the diversity and classification of plants.

Catesby, Mark (1682–1749) was an English naturalist, artist, and explorer of the New World. During his stay in Williamsburg, he met John Clayton and taught him how to preserve plants and encouraged Clayton's interest in native plants in Colonial Virginia. The two continued to correspond with each other after Catesby returned to England. Clayton also sent plant specimens to Catesby who passed these on to Linnaeus and Gronovius. He was the author and publisher of the monumental book *The Natural History of Carolina, Florida and the Bahama Islands.*
< www.catesbytrust.org >

Clayton, John (1694–1773) was born in England but traveled to the Colony of Virginia in 1715 and served as the clerk of courts for Gloucester County from 1720 until his death. Meeting and collaborating with Mark Catesby with whom he maintained a long-term friendship and correspondence encouraged his interest in plants. Clayton sent preserved plant specimens to Catesby who shared these specimens with Linnaeus and Gronovious. Some of these specimens with descriptions he sent to Gronovious formed the basis for the *Flora Virginica* (1739-1743) and second edition of *Flora Virginica* published in 1762. He also wrote his own version of a flora of Virginia that was never published and the manuscript was thought to be lost in a fire at the Gloucester County Courthouse.
< www.floraofvirginia.org/flclayton.shtml >

Colony of Virginia was a British colony established in 1607 with the settlement of Jamestown, Virginia. In 1775 it became one of the original thirteen states of the United States when independence was declared from Great Britain.

Corm is an underground storage stem similar to bulbs, tubers, and rhizomes. A corm forms from the underground swollen base of the stem with the roots forming underneath the corm. While the leaves of a plant are in sunlight and photosynthesizing to make energy-rich sugars, extra energy is stored in the corm for next year's development of the plant. Many spring ephemerals have corms or other underground storage structures.

Dichotomous key is a series of pairs or descriptions, which are used to identify a specific type of life form by making a choice between the traits and characters described in each pair or couplet.

Elaiosome is a fatty structure that is attached to the seed of a flowering plant species, including many spring ephemerals. The fats in elaiosomes attract ants, which carry the seeds away to use for an energy-rich food source and in the process help disperse the seeds.

Endangered species is a species of organism that is extremely rare and is at risk of becoming extinct (no longer existing as a species).

Genus is the singular form of "genera" and is the level of classification that is below family and above species.

Gronovius, John Frederick (1686–1762) was a Dutch botanist and an important patron of Linnaeus. John Clayton sent him a list and description of Virginia plants in the 1730s, which he translated into Latin and published as the *Flora Virginica* (first edition in two volumes in 1739–43 and the second edition 1762). His son Laurens is credit for publishing the second edition after the death of his father.

Linnaeus, Carl (1707–1710) was a Swedish botanist, physician, zoologist, and author of *Species Plantarum* published in 1753 as well as many other scientific publications. He standardized the scientific naming of organisms with his introduction of binomial nomenclature. He gave scientific names to many of Virginia's native plants based on specimens sent by John Clayton to Mark Catesby, who sent these to Linnaeus. Linnaeus honored John Clayton by naming the genus of spring beauty *Claytonia* and the species *virginica* for the Colony of Virginia. < www.linnean.org >

Plant press is a structure typically made of two ventilated frames, usually 12 x 18 inches (30 X 45 cm.), within which plants may be arranged between sheets of driers/newspaper and ventilators such as cardboard, tightly strapped together with press straps, in preparation of being added to a collection or herbarium [to preserve the dried plant specimens that can be added to a plant collection or herbarium].

Rare species is a species of organism that is uncommon or rarely seen in a given area.

Spring beauty is the common name of the spring wildflower *Claytonia virginica*, which was named by Linnaeus in 1737. This species formerly was placed in the family Portulaceae but now is in the miner's lettuce family Montiaceae.

Spring ephemerals are spring wildflowers found in deciduous forests that send up their vegetative or non-reproductive parts early in spring. They soon flower, are pollinated, set fruit, and have their seeds dispersed. As the leaves on the deciduous trees appear and begin to block light to the forest floor, these wildflowers have their aboveground stems and leaves die.

Vasculum is a flattened tin cylinder with a shoulder strap used by botanists as early as the beginning of the 18th century to keep plant specimens in good condition. It would be lined with moist cloth or paper to keep plant specimens from wilting.

Twinflower
Linnaea borealis

Epilogue

The inspiration for this book came several years ago as the Foundation of the Flora of Virginia Project formed in an effort to launch a fundraising campaign to cover the necessary cost of publishing a modern flora of Virginia. I was alarmed that if it were not for the efforts of private donors to cover the cost involved; botanists and illustrators needed to produce a modern Flora that Virginia would continue to be negligent in the study of our unique botanical treasures. Why is it that botany and the understanding and protection of our wild plants are not seriously valued at our universities and within our state government? I believe it is because we do not teach children about our regional biodiversity, the importance of native plants in the ecosystem, and services they provide to our well-being, therefore we have few advocates or caretakers who are knowledgeable about the role of plants in the landscape. This story then became a vision to plant a seed, a children's book that could be a spark to teach the basics of botany, the historical context of botanical pioneers in the state of Virginia, and to highlight the role of one single native plant in the ecosystem.

Realize that when John Clayton's second edition of *Flora Virginica* was published in 1743, approximately 500 species of plants were described. Now as of November 2012, with the publication of the modern *Flora of Virginia*, 3164 species are described. I would like to highlight the significant role of the Virginia Department of Conservation and Recreation's Natural Heritage Program and the Virginia Native Plant Society for their work and efforts to conserve Virginia's rare species and habitats.

Botany was a respected science for the most noble of individuals in the 1700's. I believe that the collaboration of so many people and organizations to achieve the modern *Flora* demonstrates that the study of plant life will again regain its due stature. This book will be sold on the www.FloraforKids.org website. Proceeds from the book will be donated to the Foundation of the Flora of Virginia Project.

I want to acknowledge and thank Bland Crowder for his editorial review and for his support and encouragement to see this project into fruition. Bland Crowder is the Director of the Foundation of the Flora of Virginia Project, a non-profit organization that continues to develop ways in which the *Flora of Virginia* can be an educational tool. To learn more about the Foundation of the Flora of Virginia Project and to purchase a *Flora of Virginia* go to www.floraofvirginia.org.

Author

Susan Leopold, PhD.

Susan is an ethnobotanist and passionate defender of biodiversity. Over the past 20 years, Susan has worked extensively with indigenous peoples in Peru and Costa Rica. She is currently the Executive Director of United Plant Saver, www.unitedplantsavers.org, dedicated to medicinal plant conservation of the United States and Canada. She also serves as a board member of Botanical Dimensions and Center for Sustainable Economy.

Susan and her three children are proud members of the Patawomeck Indian Tribe of Virginia and live on a farm where they raise goats, peacocks and herbs.

Illustrator

Nicky Staunton

Nicky is a Virginia conservationist and advocate for native plants and the habitats that support them. She was the State President of the Virginia Native Plant Society for three terms. Since retiring she has become a recognized artist, mainly in pen and ink illustrations. Her work has appeared in several books and many Bulletins. She studies dry brush watercolor with Lara Call Gastinger, who is the chief illustrator of the Flora of Virginia*; is a member of the Guild of Natural Science Illustrators, American Society of Botanical Artists, and Southern Appalachian Botanical Society.*

Her botanical taxonomic skills are from studies with Marion Lobstein and were used for the plant inventory of the 500 acre Occoquan Bay NWR in Woodbridge when the land was US Army Harry Diamond Lab.